CREATION
and
EVOLUTION

compiled
and edited by
Kenneth N. Taylor

GW00382542

 Tyndale House
Publishers, Inc.
Wheaton Illinois

Aids for Bible Education
"Charleville"
Harbour Road
Dalkey, Co. Dublin, Ireland

Photo credits: American Museum of Natural History, pp. 94, 115, 117; L.M. Beidlen from Monkmeyer Press Photo Service, pp. 46, 48; Brookhaven National Laboratory, p. 30; Gerard from Monkmeyer, pp. 44, 51, 61; Gregor from Monkmeyer, p. 53; Harrison from Monkmeyer, p. 57; Gerhard Heilmann, p. 89; Carol Hilk (after Lessing), pp. 28, 39, 41; Harold M. Lambert Studios, pp. 74, 91; Lick Observatory, pp. 6, 12-13, 15, 17; New York Times, p. 37; Radio Times Hulton Picture Library, pp. 70, 103; H. Armstrong Roberts, pp. 67, 72, 81, 85, 108; William Roberts, p. 76; Tom Schmerler, p. 125; Stonnec from Monkmeyer, p. 26; United Press International, pp. 99, 104; Watson from Monkmeyer, pp. 58-59; Department of Geology—Wheaton College, p. 83.

Library of Congress Catalog Card Number 77-81902

ISBN 0-8423-0455-X

First printing, September 1977

Printed in the United States of America

Contents

1/**CREATION**

Do the heavens "declare the glory of God" as glowingly proclaimed by King David thousands of years ago? Or are they merely accident and happenstance? Your answer, whether right or wrong, can change your life.

Do you know that on a clear night
you can see only as much of the universe as an amoeba can see of the ocean
in which it drifts? All of the glory that
the human eye can take in is compa-
rable to its drop of water! Standing
at night on the beach and being able
to comprehend the vastness of the
universe is as impossible as trying to
count the grains of sand on which you
are standing.

How did all of this grandeur origi-
nate? The Bible says, "The heavens
declare the glory of God." That is,
the stars exhibit His existence and
power. But although the vastness of
the universe should impress us, somehow it doesn't. We are used to it and
just take it for granted. And though

we can understand the argument that order and design among human beings postulate an orderer and designer, yet the thought seems strange that order and design in the heavens postulate a divine Doer. Who is this One who orders and designs our universe? Do we take Him for granted, too? Or do we just ignore Him, saying He doesn't exist?

When we see a hand-crafted cabinet we know it was made by a cabinet-maker. We have no difficulty at all in assuming this even though we did not see it made. Yet when we look up into the vastness of the universe and see stars moving in orderly relationship in vast interstellar space, we shrug our shoulders and express our inability even to guess how it all came about. We speak vaguely about chance and accident, or somehow, in the back of our minds, there is the thought that God must have done it, but it is so vast a miracle that we cannot really feel it, and no one else seems to notice it either, so the matter is forgotten. If

Examining the evidence is very important, but isn't really the first item on the agenda! Although the truth will become evident if one is open-minded, almost everyone begins with an assumption for or against God, thus biasing his interpretation of the evidence.

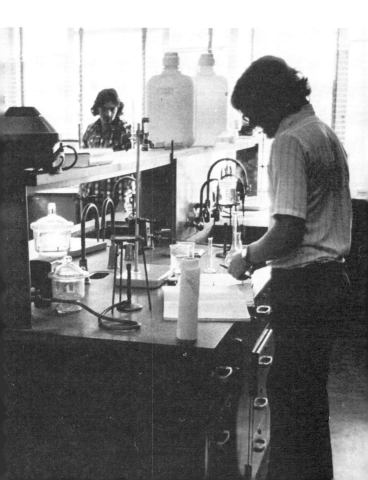

only our minds were not so dull or calloused! Would we not then recognize the universe as the handiwork of God and fall face downward in the dust in worship?

Why is the study of this subject important? Because if it is true that the heavens declare the glory of God, then here within our grasp is the key to an eternal mystery which leads us into the very presence of God. It is we who suffer loss (even though we may not be aware of it) if we ignore or turn away from the evidence for God's existence.

Our examination of the evidence begins with the macrocosm—the limitless distances of outer space. But how can we begin to take in its enormousness? The grandeur is beyond all comprehension. How shall we speak of the distance involved in one light year? For in a single year, light rushing out into space at the speed of 186,000 miles per second, travels six trillion (6,000,000,000,000) miles.

The sun is 860,000 miles across, compared with the earth's 8,000 miles—100 times as wide! (If the sun were hollow, one million earths would easily fit inside.) But another star in our galaxy, Antares, is <u>150</u> <u>million</u> miles across. Into it would fit our entire solar system out to the orbit of Mars!

This cloud of stars is part of the universe of Sagittarius. Our own universe is 100,000 light years across and contains an estimated 100 billion other stars besides our sun. But ours is only one of a billion known universes.

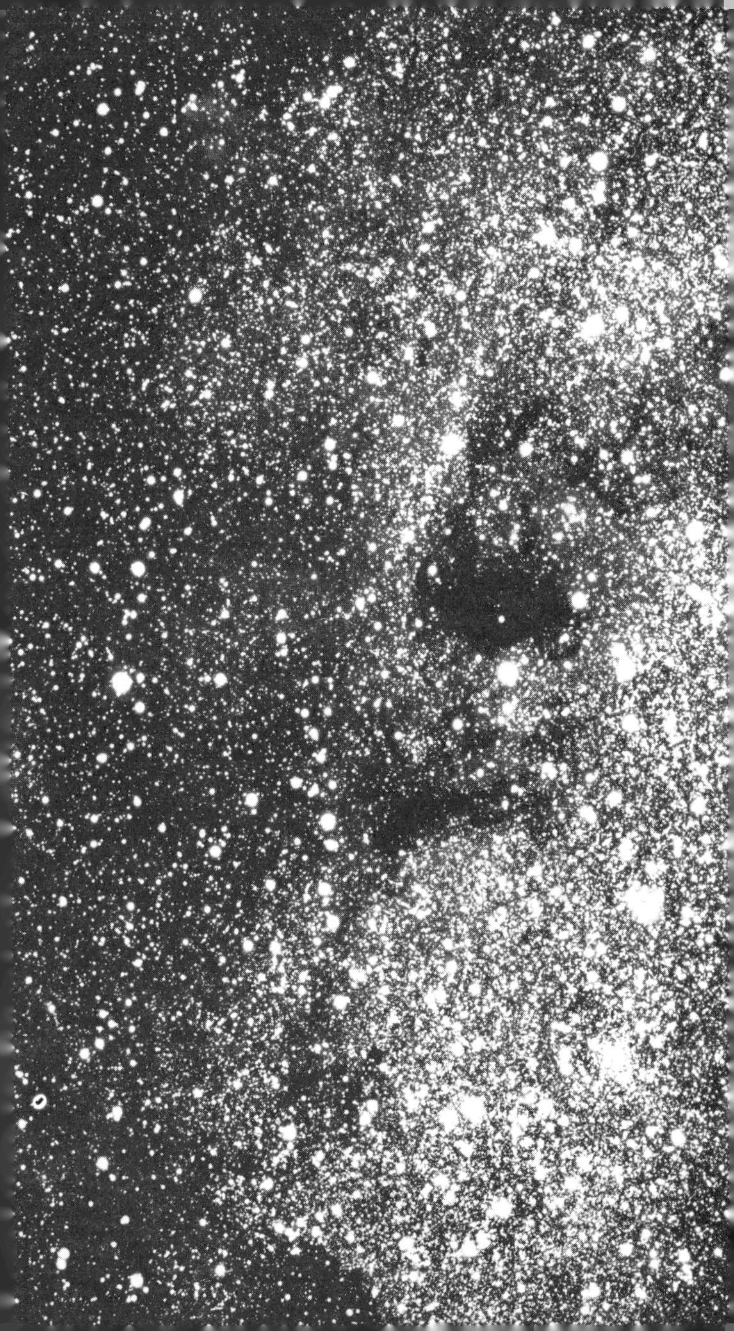

Can you begin to see the glory of God and how the heavens show it to us?

This solar system of ours—the sun (which is a star) and its satellites—is part of a huge galaxy of 100 <u>billion</u> other stars formed into a giant pinwheel in space. Our solar system is out toward the edge of the pinwheel. It would take 100,000 years for light to travel from one end to the other, or 20,000 years through the thickness at the center. This giant pinwheel revolves "slowly" through space, with our solar system located in the part of the pinwheel traveling at the rate of 136 miles per second. Yet it takes 200 million years for the pinwheel to make one complete rotation.

When we look up at the Milky Way

we are looking into the pinwheel. The far fewer stars we see beyond the edges of the Milky Way are because we are no longer looking into our galaxy, but away from it, toward others.

Our earth is in a pinwheel galaxy somewhat like this one. We are located in the part of the pinwheel rotating at a speed of 136 miles per second. 200,000,000 years are required for one complete rotation.

But now think of this! Our pinwheel galaxy of billions of stars,
100,000 light years across, <u>is only one
of a billion other</u> galaxies, some of them
pinwheels, visible through telescopes
from the earth! The closest of these
other universes is 200,000 light years
away. It is estimated that in addition
to the billion universes we can see,
there are other billions beyond our
sight and comprehension, each containing hundreds of millions or billions
of sun-sized stars. Sir James Jeans
has suggested that the number of stars
in all the universes may equal the
number of grains of sand lying on all
the seashores of the world. Some of
these other constellations of stars out
at the edge of space (by which we mean
the farthest visible to our observation
systems) are apparently moving away
at a rate of millions of miles an hour,
almost at the speed of light.

And far out beyond the farthest
reaches of our imagination, who is to
know whether space goes on and on a

billion times as far? And where does
it end? Are there *no* <u>boundaries?</u> And
did all this happen by chance?

Limitless space? *Never* ending? *No* boundaries? All the evidence points in that direction.

But though the stars number in the billions, yet space is almost empty! If our sun were the size of the dot over this "i," the nearest star in our pin-wheel galaxy would be a dot ten miles away and other stars would be up to the size of a dime, hundreds and thousands of miles distant.

And now the question is, are these vastnesses of space—these never-ending distances, these eternities of time—are they mere chance formations or are they in a majestic order created by God? In other words, are they meaningless, or full evidence that confirms the existence of God?

The answer is extremely important. For if God exists, then you and I have a responsibility to know about Him—and to love and worship Him. But if the universe originated through chance and there is no Creator, then nothing in this universe or in our lives has purpose. All is accident and chance and we can "eat, drink and be merry," for there is no tomorrow and today is not valid. We are lost in the vast reaches of eternity, on a tiny planet, among a trillion stars. (What horror if God should lose track of the earth—a grain of sand lost upon the seashores of the world!)

We are not playing games. These incredible facts are real. Yet how many people blithely ignore them and go about their business, each to his own little way as though nothing else mattered—neither time, nor eternity, nor space, nor a billion billion stars a million times as big as the earth, nor never-ending space, nor God.

Yet the Bible says that God created it all: "By the Word of God the heavens were of old," and "Without Him nothing was created," and "In the beginning God created the heavens and the earth."

How do people come to the conclusion that the universes are mere accident and chance despite the enormous complexity of their movements, balance and interaction? Such people begin with the presupposition that

Good minds, good teachers, good textbooks—all are important. But the decisive factor of truth may never be ours unless we are willing to recognize the fact of God behind all nature. If He is there, and we ignore Him, our system is false.

there is no God. Then they say, "Since there is no God, then everything must be by accident and chance." This is what is known as circular reasoning and begging the question. Is it not more reasonable to declare accident and chance unlikely or impossible, and to presuppose God?

21

The microscopic world becomes increasingly complex as it is magnified. The most powerful student microscope barely "scratches the surface." Much of what we know about the invisible worlds of molecules, atoms, and electrons is by theory and calculation.

22   *The Big Bang?*

Not many theories have been suggested as to how the universe began. The "big bang" theory is currently the most popular. This theory postulates that sometime in the inconceivable past, billions and billions of years ago, vast fields of gas (where it came from, no one can guess) gradually condensed and finally became incandescent. Then, perhaps set off by heat, a blast occurred that shot the galaxies into space at inconceivable speeds. These whirling masses of molten universes have been rushing out through eternity ever since—for possibly five billion years. The galaxies farthest out, billions of light years away, move at

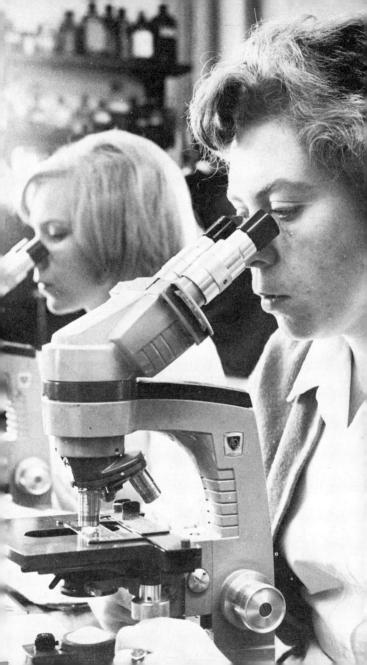

the greatest speed, at almost the speed
of light.

The other main theory has been
that of Fred Hoyle: the steady-state
idea. He postulates that matter is
constantly developing from nothing,
continues for a few billions of years,
and then disappears again into nothing.
But in recent months Dr. Hoyle has
himself begun to doubt the validity
of this idea.

How it all came about—whether by
a "steady-state" or a "big bang," or an
instantaneous calling into existence—
whatever the means, the incredible
greatness of Someone's power is evident. Do not hesitate to let the wonder
of it fill you with praise and admiration
and worship. That we don't know how
it was done is irrelevant. It is the fact
that matters. Yet some people find
this fact too incredible to believe. How
would it be conceivable that there could
be a billion universes in infinite space
with no ultimate walls? Impossible, yes,
but true. There it is—right up above

us in the sky. If this is the work of God, He must be great indeed.

Human bloodcells

And now let us turn to the other side of the universe, to the world of plants and animals and cells, of molecules, atoms, protons, neutrons, and electrons—and who knows what else may be discovered in even tinier dimensions— contrasting immensely with the enormous worlds overhead.

In order for man to see these microcosmic elements they must be magnified by powerful microscopes just as the macrocosm around and beyond the earth must be magnified and brought closer by powerful telescopes. Although the contrast in the size of these two worlds is indeed great, many similarities exist between them.

The atom. Electrons in random orbits circle the nucleus billions of times in millionths of seconds!

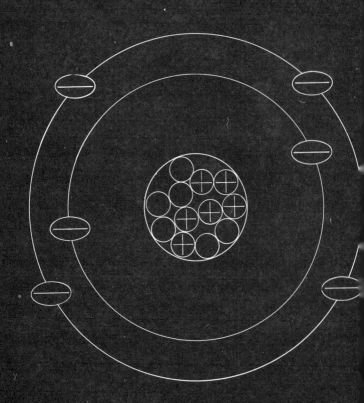

The electrons surrounding the nucleus
in an atom are worlds apart, as far
apart in proportion to their size as the
planets are from the sun. And the electrons move in orbits (though random)
just as the planets do. Electrons
whirl around the atom's nucleus at
fantastic speeds. They complete
billions of trips around the nucleus in
millionths of a second!

Amazing, incredible, glorious,
fantastic.

Chance?

Accident?

Everything you see is made up of
atoms—your chair, the floor, the wall,
this book, the air you breathe, your
body. Each consists of billions and
billions of atoms so tiny no one can see
them, yet so vast that there is a universe between their component parts.

Since it is true that atoms are mostly
empty space dotted occasionally by
weightless electrons with great distances
between, it is also true that the chair
you are sitting on is mostly nothing-

A high-sensitivity rare-gas mass spectrometer at Brookhaven National Laboratory. It is capable of detecting as little as one-billionth of a cubic millimeter of gas. It is used in the study of the age of meteorites.

ness, held together by the force of whirling electrons moving so rapidly that they cannot be crushed. No wonder the Bible says that the things that appear are made of things not seen.

The Origin & Complexity of Life

Where did life come from? Was it created by God as the Bible tells us, or did it just happen? Let's look at some of the reasons why it seems unlikely that life began by chance. First, consider what the bio-chemists are doing in the laboratory these days in their efforts to create life in a test tube.

They have made remarkable strides in producing tiny molecules. They do this by passing electric charges through various chemical solutions. The resulting molecules are far too small and simple to carry life, but they are a step, and someday perhaps they will be able to do it.

(A much asked question in science

classrooms these days is: "If scientists can create living matter in a test tube, doesn't this rule out the existence of God?" Why need it rule out any such thing? Who started this strange question, and who checked its logic? For example: If my father builds a house, and I watch how he does it, and do some experimentation and research and build one myself, does this prove that my father doesn't exist? The logic is the same.)

And now we come to the structure of the simplest form of living matter, the cell (see illustration). As you can see, it is not, after all, very simple! For a cell to be formed by chance, millions of complex protein molecules would have to "spontaneously generate" simultaneously and then by strict accident and chance join together and form into various parts of a protozoa or other single-celled unit, including such parts as the extremely complicated chromosomes and genes: otherwise, the cell could not duplicate and redup-

A diagram of a "simple" cell! Incredibly complex "machines," far more sophisticated than the most advanced data-processing mechanisms, are in every cell. In the nucleus are the chromosomes containing the coded DNA that specifies every aspect of the growth of the body. A teaspoonful of DNA is estimated to have an information capacity equal to a modern computer with a volume of 100 cubic miles!

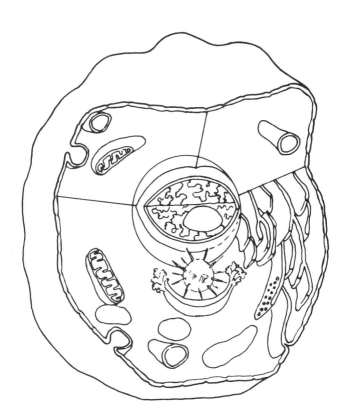

licate itself over and over again billions of times to form a human body or other form of living matter. Notice that this division of cells is not the least haphazard: an embryo within its mother is not a shapeless mass, but a shaped foetus; and the foetus continues to change. The ever-dividing cells form themselves into the digestive system, respiratory, circulatory, and so on, until the original cell has become the incredible complexity of a man or a woman.

Then, at a certain point, there are no further significant changes in the unborn child except in size. How do the cells know when to differentiate into organs and when to stop growing? Who made the chromosomes and genes that seem to be the controls? And how do these controls work?

Let me further describe how complex a cell is; for my purpose is to try to convince the open-minded that no amount of chance would form a cell, and that a Creator is required.

How new cells are formed: By the process of mitosis, the chromosomes divide and pass across spindles to form the nucleus of two new cells. Each cell of the body thus contains the identical code of every other cell in that body. This basic pattern of reproduction is not the work of chance mutations, but of the Creator.

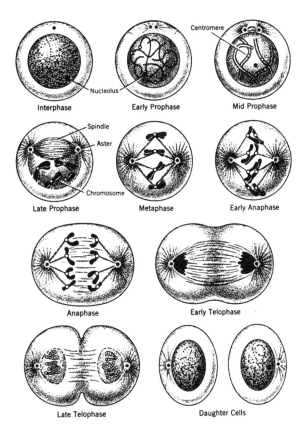

Interphase

Early Prophase

Mid Prophase

Late Prophase

Metaphase

Early Anaphase

Anaphase

Early Telophase

Late Telophase

Daughter Cells

A model of the complicated DNA molecule, which is part of every cell and is the machine that manufactures proteins from raw materials.

Have you been reading about DNA? (The letters stand for deoxyribonucleic acid.) DNA is the "machine" in the chromosome of the cell that duplicates the chromosome, including all the hereditary features to be passed on to the daughter cell whenever cell division occurs. That is, the DNA controls the color of your eyes, the shape of your body, your personality, and a million other facts about you that are different from a dog or horse, or from another person. And the DNA in each cell of the dog or horse gives those animals their special characteristics that cause them to be what they are and not a different sort of being.

How does DNA duplicate its information and how does it pass it on to its daughter cells?

DNA is a complex molecule built in the form of a double spiral, or helix (see illustration). The two backbones

are long chains of alternating sugar and phosphate groups. These two backbones are joined by four kinds of connecting rods, A, T, G and C, appearing in various sequences. In fact, so many sequences are possible that it is estimated that there are at least 4 or 5 <u>billion</u> different kinds of DNA molecule combinations in the forty-six chromosomes of man, each controlling one of his features. (It's the same idea as in a dictionary— thousands of different words and meanings come from twenty-six letters in various sequences.) It is estimated that if one could collect a teaspoon of DNA, it would have an information capacity equal to a modern computer with a volume of 100 cubic miles!

How does DNA get its message across? That is, how does the code it contains direct the various cellular processes? The illustration (page 41) makes it clear: The large mass in the illustration is a ribosome, a sort of protein factory floating in the cell. The "master plan" DNA in the chromosome (not

DNA again. This diagram
shows how the
DNA molecule splits
apart during cell
division.

shown) has already made a copy of itself on a messenger strand of RNA. This strand is open at one side so that the "manufacturing mechanism" can mesh into it (see illustration). The RNA messenger enters the ribosome like a ticker tape. Another kind of RNA, called transfer RNA, acts like the machines in the factory, putting the raw materials together in a special pattern as they follow the ticker tape directions! One end of the transfer molecule has bases that exactly match one of the three-letter words on the messenger. The other end will match only one certain kind of amino acid, a raw material of protein. Each transfer molecule, therefore, can bring in its particular amino acid only when the messenger calls for it. The amino acid molecules link and react with each other to form a particular chain of protein, one of 100,000 varieties, just as we can spell thousands of words with twenty-six letters. But "somebody" has to know how to spell!

This diagram shows how the DNA code within every cell directs the manufacture of proteins from its "tape," which is inherited and reproduced generation after generation.

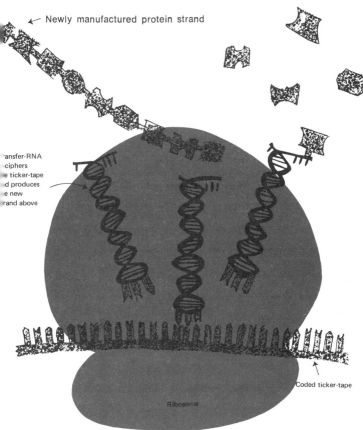

← Newly manufactured protein strand

ansfer-RNA
ciphers
e ticker-tape
d produces
e new
rand above

Coded ticker-tape

Ribosome

The protein factory in every cell

As the messenger strand emerges from the other side of the "factory," it feeds into other ribosomes to direct the formation of more protein. In this way thousands of protein strips are produced in a single minute from just one strip of messenger RNA.

We have only given a summary of one part of one process in the life of a cell. We have not discussed countless other processes, such as how food comes and goes, how enzymes act as "traffic directors" to stop and start each of the operations, how a single cell or group of cells initiates the process of differentiation. Why doesn't an embryo which begins growth by single cell division simply become an adult-size embryo, rather than an organism with specialized tissues and organs? These life processes, according to some theories, arose by the chance combination of raw materials, without any participation by God. What is your opinion?

The study of instinct is not only fascinating, but wonderfully instructive concerning the relationship of God to His creation.

Now we turn to another area of life to observe amazing things that could not "just happen" by chance in nature. Take a look at the world of the ants.

Ants live in colonies, for their duties and talents are so specific and varied that it is a case of "all or none." They must have each other's support and help—or perish. Some collect food and bring it back to the nest. Others enlarge the nest-home or keep the rooms clean. Still others take care of the queens and the growing ants. Nurse-ants have the special duty of cleaning and feeding the larvae. Also, they move the larvae to different parts of the nest if the nursery becomes too wet or cold.

Where does the instinct come from,

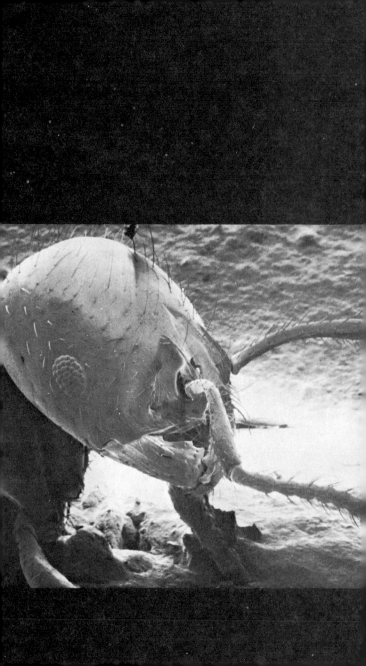

The amazing societies of ants and bees cannot reasonably be attributed to natural processes such as mutations and natural selection. It is much more reasonable to say that their instincts are God-given.

and what is instinct, that one ant should **47** perform one task and another a different task, and not all choose the same task?

Some ants are slave keepers and cannot live without slaves because their jaws are so long and sharp and curved that they cannot dig nests or feed themselves. Without their slave ants to help them, they die. How did this clumsy mutation survive? How did this "negative survival of the fittest" occur? It was apparently not by accident and chance!

The worker ant's responsibility is to provide food, mainly wood or grass. Soldiers protect the columns of ants who go out at night to cut grass in quarter inch lengths and bring it back and stack it in storage chambers. These soldiers make up about five percent of the group. They have protruding mandibles, with a nozzle in the head for squirting chemicals at the enemy; they throw out a gummy substance that tangles the legs and antennae of the enemy and irritates its body. If their

nest is attacked, the workers begin
sealing off tunnels leading to the outside, cementing the new fragments of
soil into place with feces.

When the queen ant becomes
sterile, (at that time and not before),
one of the young females starts developing the ability to mate and take over
egg production. To say that this is all
"by instinct" is apparently true, but
who gave the instinct?

Perhaps these facts are meaningful.
Perhaps these things did not just happen.
Perhaps God with supreme intelligence
and power designed and created ants,
bees, homing pigeons and all the other
amazing assortment of creatures exhibit-
ing such common instincts as mating,
running away, migrating, etc.

The leaf-cutter ants don't eat leaves, but fungus, which is their only food. These ants use the leaves to make subterranean compost piles where their fungus farms flourish. When a virgin queen of the leaf cutters sets out on her mating flight, never to return to her old colony, she always takes with

her a small piece of fungus from the colony's fungus patch. She carries this with her in a special little pocket beneath her mouth. After she has mated and shut herself up in an underground cell, she at once starts a fungus garden, planting the piece of fungus she has brought with her, fertilizing it with her own excrement, thus growing enough food to nourish herself and the first few tiny workers that hatch from her eggs. Since fungus is the only food of this kind of ant, she would starve if she forgot to carry some with her. The habits of this queen ant offer one of the many interesting and questionable points about the evolutionary theory of natural selection. For if the

theory is true, the idea of carrying along the fungus must, of necessity, have mutated simultaneously with the idea of the mating flight. It seems unlikely that such a complex set of ideas could evolve. It seems more likely that her habits arise from a God-given instinct. Incidentally, the pocket in the queen's mouth for carrying the fungus would have had to mutate simultaneously with the idea. This idea of carrying along a piece of fungus is indeed very foresighted of the queen ant, since she has never been away from the nest and has to think through in advance the strange conditions that are before her.

Ants make good dairy farmers! A great many different ants are interested in the small sap suckers (aphids) because of the sweet liquids they excrete; these are their "cows"! Foraging ants frequently come across groups of these aphids that have settled on some leaf or juicy shoot and sunk their stylets into the tissues of the plant. As these

insects feed, they excrete waste products which either fall to the ground or accumulate on the leaves as a sticky coating. This is known as honey-dew, and the ants, with their liking for sweet substances, are very eager to gather it up. When an ant strokes an aphid with its feet and antennae the aphid exudes a drop of this lovely liquor. The more the aphid is stroked, the more honey-dew it produces and the better it thrives, for its rate of metabolism is increased. It grows faster and reproduces better when it is kept clean by the ants.

Wood ants keep herds of aphids in underground chambers where they are fed on roots. Many ants build special shelters for their "herds." Sometimes these "barns" are made of fine grains of soil cemented together. The tropical weaver ants build special silk tents for their "cattle." It was a great day for the aphids when, somewhere down the line, an ant hatched with this interesting talent of being able to make "silk" tents for aphids!

In addition to using aphids to milk honey-dew, some varieties of ants also slaughter some of their "cattle" for meat.

Ants, like bees, when a large amount of food has been discovered, perform a kind of dance to tell each other about the direction and the distance of the good source of food! Then their companions set out at once and go directly to it without hesitation!

Aphid milking is one of the choice tasks. Specialists remain in charge for many weeks before taking a furlough. "Cowboy" ants stand guard to protect the herd and to keep it from straying. Dairymaids do the milking while special "tank-car" ants (unrefrigerated, though!) carry home the honey-dew from the "barn" by filling their crops with it and disgorging it again at the nest.

The tropical weaver ants have "learned" to use the silk produced by their larvae. The mother holds her

larva in her jaws, and using it like a
shuttle, passes it back and forth between
the edges of several leaves, thus making
a network of tiny silk threads, which
draws the leaves together and forms a
nest with roof and walls.

Ants are usually vicious fighters
and if their homes are attacked by
other ants and there is no nearby nest
to move into, they will defend themselves with great vigor. But if there are
other nests, they will abandon their
home to the invader and move. Other
ants from nearby ant hills will often
come to the assistance of their friends
in trouble, helping the refugees move
their broods to safety. And who decides whether to attack and when to
defend or whether to move? If the
raided nest is small and weak the whole
military operation may be completed
in a matter of hours but sometimes the
raiders return each afternoon, day after
day, until all defense is broken down
and the attackers take over the nest.

A single hive of one kind of bee can swarm as many as thirty times in one season. The bees "decide" in advance how many young queens to raise, often depending on the severity of the preceding winter.

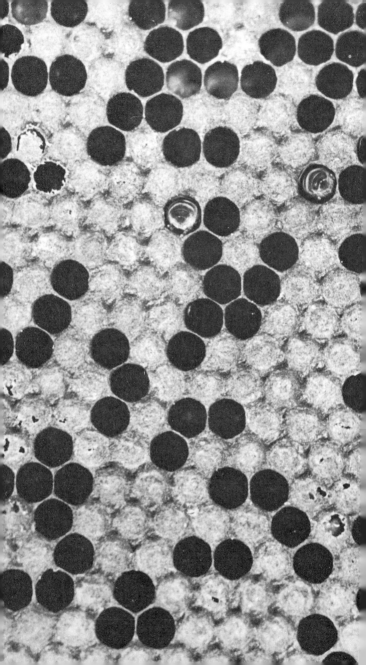

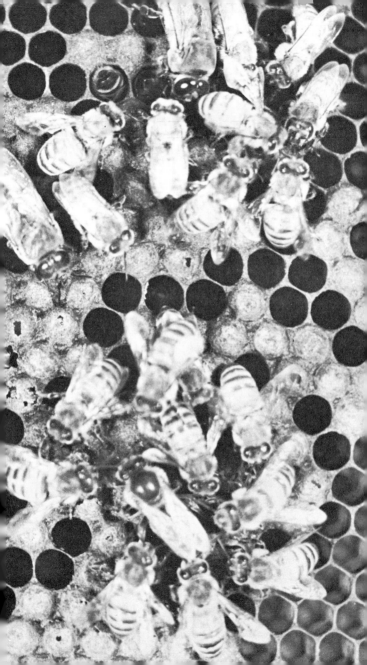

Bees, like ants, work together. Their instinct for division of labor inside and outside the hive is difficult to understand and wonderful to see.

A single hive of one kind of bee can swarm as many as thirty times in one season, sending out thirty new colonies. After an especially severe winter the bees "decide" to raise more young queens in order to replace colonies that may have died out under the severe conditions. These swarms are possible only if there is an extra queen to lead them. The new queens (usually only as many of them develop as are needed for the available swarms) are developed by the workers by overfeeding the larvae. Who makes this decision as to how many swarms there are going to be, and how many queens should be developed?

The queen ant, as soon as her mating flight takes place, breaks off her wings because she realizes (!) that she will never need them again throughout

her long life ahead; they would only
be in the way.

These, then, are some of the wonders
of creation, showing the interesting
ways God has taught the animal world
to care for itself; ways that are similar
to human ways, but apparently without
intelligence and thought.

The vast worlds above us and the
incredible worlds at our feet are evidences to the open-minded of inconceivable intelligence and power. But
without open-mindedness, the strongest
evidence will fail, for "there are none
so blind as those who will not see."
To those willing to believe, the glory
of the heavens and the infinite grandeur
and order of the atomic world, and the
strange, improbable world of nature,
all point to the infinite God of creation.
How else can these facts be explained?
And it is this same all-powerful Creator
of ours who loves us and longs for us
to love Him with our entire lives—
spirit, souls and bodies. What is your
reply to Him?

2/EVOLUTION

Why all the hangup about evolution?

Why the outcry against it by many Christian parents, and the outcry of many scientists against anyone who doesn't believe it?

What difference does it make whether life "just happened" and then slowly evolved into man, or whether God—a real, personal, all-powerful and all-intelligent God—made everything there is: not "man from monkey" but man and monkey as special creatures from the hand of God, in just about their present shape and form?

Well, there are two main reasons for the controversy.

First, the Christian is ticked off by the strong assertion of some evolutionists that the "fact of evolution" has freed them from "superstition" about the existence of God. So the Christian, with his strong belief in God, naturally wants to believe in something else than evolution if atheism goes along with it.

Second, there is the question of the reliability of the Bible. For the Bible, on which the Christian student rests his faith for both this life and the life to come, seems to teach the instant creation of the man Adam in the Garden of Eden. So if it didn't happen that way, then is the Bible trustworthy? And if not trustworthy in regard to this pre-history, how can it be the Word of God? If the Bible is not true about creation, then what parts of Scripture are true, and how can one decide which are true and which are not?

The purpose of this booklet is to discuss evolution from the viewpoint of the creationist—the person who be-lieves that God created Adam as a full-grown man. The book will show why evolution remains merely a theory, rather than being a fact on which a

case against the Bible can be built.
Then it will show why the theory does
not seem to the creationist to be true,
probable, or even possible—despite the
fact that almost all biology teachers
and textbooks teach it (so we have a
lot of explaining to do!).

First of all, though, it should be
pointed out that disproving evolution
or proving creation (or God) is impossible, for there is no solid biological
evidence for evolution, and it is difficult to disprove what can't be proved.
There are facts, such as human fossils
perhaps a million years old. But there
are two ways to interpret the facts—the
evolutionary interpretation and the
creationist interpretation. And neither
can claim that the facts prove one
position or the other. How could they?
For facts that happened apparently
millions of years ago are hard to check.
About all that can be stated with cer-
tainty is that the evolutionist thinks
his interpretation of the facts is good,
and the creationist thinks his is better.

In other words, one's interpretation depends on one's presuppositions. If one's presupposition is that God created without evolution, then a creationist system is the inevitable result, and all problems resulting from that viewpoint are cleared away by various suggested solutions. And if one assumes that all things emerged in evolution, then the very difficult problems stemming from that point of view are similarly cleared away by the evolutionist as best he can.

Now presuppositions aren't necessarily bad! All of us believe many things we can't *prove,* and this can frequently lead to truth. The creationist, too, has presuppositions: for example, the assumption that there is a God. And we think we have good evidence— answered prayer, changed lives, the universe all around us, fulfilled prophecy, etc. But these are not scientific proofs in the usual sense of the word; and when science ignores or attempts to "disprove" God, we still believe in Him, for He is our presupposition.

Similarly, the evolutionist bases his theory on presuppositions. But in his case the situation is less reasonable, for he is dealing with a scientific theory and he should be able to marshal adequate scientific facts to support his theory. As this paper will try to show, he does not do this, but he nevertheless insists on holding to evolution as an established fact. Instead, he should frankly state that much of the evidence is lacking and that his belief is based primarily on presuppositions; and of course, some do.

So the argument can't be won on the basis of the facts. Neither the evolutionist nor the creationist comes to his conclusion from evidence acceptable under the usual laws of scientific proof. So there is considerable difference here between weight of opinion and weight of evidence!

Charles Darwin (1809-1882).

A Bit of History

Now for a bit of history before getting into the arguments pro and con. About a hundred years ago, Charles Darwin began realizing that there are variations in animals of the same kind. Notice, as an example of this, the different races of people, and different kinds of dogs ranging in size from Pekinese to Great Dane; yet all men are part of the same human race, and all the varieties of dogs are dogs. Darwin came to the conclusion that variations occur in all animals and plants, and that the animal with a useful variation survives longer and therefore has more time to have more offspring. Often, some of these offspring will have the same variation, and the

This is not evolution. Different breeds of dogs can be developed from a common ancestor, but all are still dogs.

variation may become increasingly prominent as generations roll by. Finally, generations later, the variation will be normal rather than being the exception. This is known as variation by natural selection.

The polar bear may be an example of natural selection. Of the various-colored bears that may once have inhabited the arctic, white bears are better protected from their enemies because they can't be seen against the snow. This could mean that white bears often lived longer than other colors of bears in the arctic and so were able to produce more offspring, until finally whiteness became standard.

But what has this to do with evolution? Well, everything! For it is not

claimed that one animal evolves suddenly into another—a reptile suddenly
hatching into a bird, for example—but
this is a process, says the evolutionist,
that takes place over millions of years
by very minute changes, generation
after generation.

From observing variations that had
become standard, Darwin came to the
conclusion that all life began from
some simple form. He reasoned that
this simple form reproduced with slight
variations that eventually became
permanent. There were slight additional changes in later generations until
finally the simple form of life, after a
long process of constant mutations, and
after hundreds of millions of years,
became man.

The creationist of course does not
believe this. He believes that God
created man separately, and that man
did not evolve from a simple form
after ages of changes. The creationist
believes that although there are variations among dogs, there are no variations beyond dogness. The Great Dane
will not become, after millions of years,
anything but a dog.

The reasons for this belief of the
creationist are essentially two: The
laws of genetics as we know them do
not include this possibility; and the
fossils do not confirm that this has
happened in the past.

What Do the Chromosomes
Say About Evolution?

The evolutionist says that the genes
in the chromosomes of the early forms
of life changed dramatically as the eons
rolled along, building up additional
codes that caused variations in the
cells they controlled. These variations
were so radical that ultimately the
first single cell became a complex
modern organism such as man.

Well, it is hard to disprove such a
theory. All we can say is that in all
the cases ever observed by anyone, this
has not happened. Fifty years of
fruit-fly experiments, equivalent to
1000 human generations, give no
different reading. The Pekinese is still
a dog and never develops into some-

thing else. Whether, given millions of years, some unknown law of heredity would be revealed, we do not know. Unless this happened, the creationist sees no way to save the case for evolution.

What Do the Fossils Prove?

Why does the evolutionist believe that the bear's ancestors (and the ancestors of all mammals, including man, and of all birds) were reptiles?

These days he usually begins his proof with the fossil record. (Thirty years ago he probably would have begun with the argument that "ontogeny recapitulates phylogeny"—that is, that there are supposed similarities between, for instance, the adult reptile and the mammal embryo, and that these are evidences for evolution. However, present day evolutionists have abandoned this argument.)

So what does the fossil record show? It seems to show that in the earliest layers of the earth there were smaller forms of life than those living in later layers.

Does the fossil record show that the ancestor of man is a reptile? No, it doesn't. Does it show that God created reptiles and bears as distinct creations? No, it doesn't show that either.

Why then does the evolutionist put so much value on the fossil evidence? How does he use it to support his views?

It is mostly a matter of logic. The early, smaller fossils preceded the later, larger ones, and so it is only reasonable, he says, to believe that the one came from the other. He cannot prove it, but it satisfies him.

The creationist, however, is far from satisfied with trying to settle the

The lower jaw of the Perry mastodon, discovered in an ancient swamp by excavators in October 1963, at Glen Ellyn, Illinois. Sixty percent of the bones were recovered, including this jaw, the skull and both tusks. The age is estimated at 11,000 years. Mastodons became extinct 8,000 years ago. These animals were similar to present day elephants (except in size!) but were not ancestral to them.

matter by evolutionary logic. For the evolutionist bases his argument on a presupposition that God didn't separately create the later forms of life. But this is exactly what the creationist thinks happened. (More accurately, *many* creationists think this. Other creationists doubt the validity of the fossil record altogether, and believe that all these kinds of life, both small and large, were created *simultaneously.*)

The creationist goes much further than objecting to the presupposition on which the evolutionist decides the matter. The creationist points out that the various kinds of animals appear suddenly without fossil links to connect them with earlier, sometimes smaller, forms.

84   The horse is a good example of this, for it is used as a standard argument by the evolutionist to show how evolution works. The facts, to which all evolutionists and most creationists can agree, are that the horse of 60,000,000 years ago was smaller and 4-toed as compared with today's 1-toed species. But as to the interpretation of these facts, the evolutionist sees in these changes not only the outworking of natural selection, but also proof that if we could only find the earlier fossils, they would show a continual, further gradation downward into some entirely different, dog-like or racoon-like animal. The only trouble is, the fossil record ends exactly where it does, with the 4-toed horse. And why does the

record stop there? Probably because that is where the horse began, for that is when God created it.

Does this horse sequence "prove" evolution? Can it not just as well "prove" creation? All available *evidence* points to a *sudden beginning* of the horse, small and 4-toed, containing genetic potential for differentiation into a 1-toed horse; also genetic potential for differences in sizes such as we have today, for instance, between the little Shetland pony and the huge Clydesdale draft horse.

Why do we not find, in the rich masses of fossils now available, a long series of fossils in known time-order, gradually becoming a horse? And since we do not find this series, is it because it didn't happen that way?

The evolutionist takes note of this and is much puzzled by the problem of missing links, but assumes that someday the gaps will be filled with the millions of links now missing.

But there is a further problem in the fossil record. Every major group (phylum) except perhaps the vertebrates is represented in the *lowest* fossil-bearing layers of the Cambrian, 600,000,000 years ago, with no evidence at all that one developed from the other. These facts are of course fully known and recognized by any competent evolutionist, but are laid aside with the hope that someday the puzzle will be explained.

But are there *no* links between the various major groups of animals in the fossil record, between the reptiles and the birds, for instance? Only a few possibilities have been discovered thus far in the fossil record. But with several millions of different kinds of fossil plants and animals known, it is only reasonable to expect that *millions* of links would have been found by now. There should be an infinite gradation from one kind of animal into another, but such is not the case.

Let us look at a creature claimed by the evolutionist to be one of these links: the Archaeornis. The evolutionist believes this "half bird, half beast" to be a link between the reptiles and the birds. It looks a bit like both (see illustration).

But the creationist sees no reason to think that it is. In the first place, if this were a link, would there not be many, many more links, showing a gradual development of bird-like structures and a gradual dropping away of the reptile-like features? (Or visa-versa —whether the Archaeornis was going up or down the supposed evolutionary ladder is indeterminable unless further links are located and dated.) But paleontologists have not discovered

The Archaeornis, from the Upper Jurassic, 180,000,000 years ago.

As all buildings use similar construction units, so also the Creator has used the cell and its fantastically complex components as the basic unit of all living material.

these other elements in the supposed chain. So the creationist's conclusion is that the further evidence probably doesn't exist, and that the Archaeornis is not evidence for evolution.

But why the similarities to reptile and bird in the same animal? Does not this *seem* to prove the evolutionist's hypothesis? No, replies the creationist, not necessarily, and for this reason: Was God as Creator required to make every kind of animal different from all others? Could He not use a basic pattern and then make some variations from it? To use an inadequate present-day analogy, if a carpenter builds a cottage and a mansion, is it illogical for him to build medium-income housing too, using basic rooms in all? Let

Cave drawings from 15,000 to 30,000 years ago. Do these prove evolution? No more than the present day American Indians do. Throughout history there have been cave-dwellers side by side with agriculturally-oriented peoples living in their tents, houses and other shelters.

Black Bull (detail of a cave painting).
c 15,000-10,000 B.C. Lascaux (Dordogue), France.

92   God be God and if He has occasionally created a creature that has features of two different groups, this only indicates to the creationist that God, with His infinite varieties, created it that way. But to the evolutionist, Archaeornis is one of the few links he has found of the millions needed to support his theory.

What About Good
Old Neanderthal?

Now we turn to the interesting questions about fossil man and cave drawings. How old are these and where do they fit into the creationist's scheme of things?

As to the age of man, the creationist and the evolutionist have the

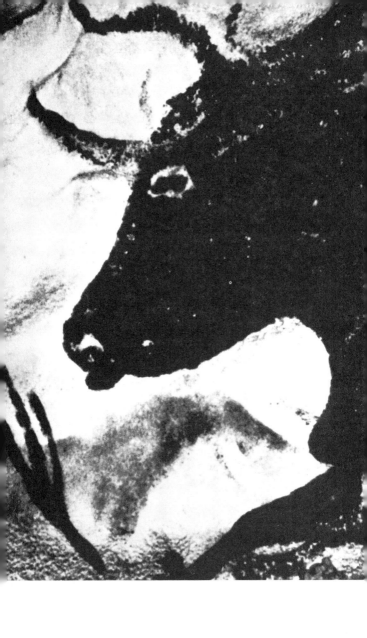

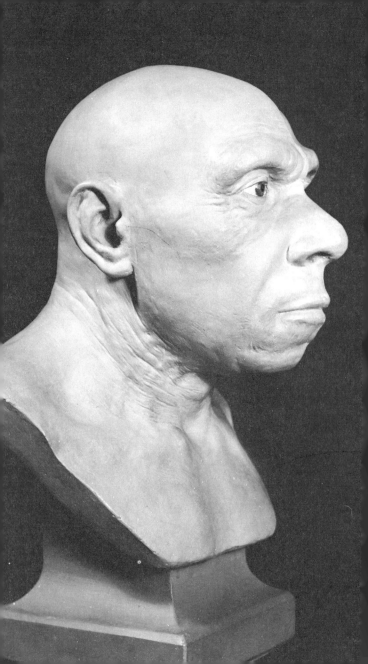

same data, with estimates ranging up
to 1,750,000 years if some recent South
African finds are actually those of man
and not of an ape. (There is no agree-
ment yet by anthropologists on this
interesting point.) So from the creationist viewpoint at this time, the picture is this: All fossil men and women
are descendants of Adam and Eve, who
were created directly by God, so Adam
and Eve are older than the earliest
human fossils.

95

The Bible gives no evidence upon
which we can draw to determine the
time of Adam's creation. Genealogi-
cal tables of the Bible, in the Hebrew
usage, list only representative ancestors
and are not a complete listing. So the
Bible permits millions of years as
easily as thousands, and is not helpful
in deciding this question as to when
Adam was created.

And what of the cave men, the
picture painters of ancient Europe? Do
they prove evolution? No, no more

than the American Indian cliff-dwellers
of Arizona. Today, and in every age,
some have lived in caves. This does
not make them any less human, and
adds no weight to the evolutionist's
claim that modern man has evolved
from these earlier men.

But is it not true that the Heidel-
berg man, the Neanderthal, etc. are
stooped and ape-like in appearance,
and therefore are obviously in the line
of ascent from molecule to man? No,
no more than differences in facial feature and body structure of modern man
prove this. The tall Watusi African,
the pigmy, the flat-nosed Asiatic, the
Negro with his distinctive features—all
are variations within the human family.
So also ancient men probably varied
from each other and from us today.
But this is not evidence for evolution
any more than is the 4-toed horse.
Man is still man.

(By the way, recently a "new look"
has been given to some of the ancient
human fossils, and they are being pic-

tured as less stooped and "ape-like" than the earlier reconstruction efforts, influenced as they no doubt were by the assumption that early man would be ape-like in appearance.)

Did Life Evolve from Inorganic Matter?

Can life develop from non-living material? If not, evolution gets off to a shaky start. It was once thought that flies in manure piles were examples of spontaneous generation. Then came Louis Pasteur who showed that the flies came from eggs that had been laid there by other flies. The idea of spontaneous generation had to be abandoned. However, in recent years the theoretical possibility of developing lifelike (though not necessarily living) molecules has become an apparent possibility. Although the idea that a rock might become an animal is manifestly absurd, it doesn't seem so preposterous when the rock is reduced in thought to primordial ooze.

Professor A.I. Oparin, author of *The Origin of Life* (New York: Mac-Millan Co., 1938), believed that simple, organic, lifelike compounds such as hydro-carbons might arise spontaneously under careful laboratory conditions.

Building upon Oparin's work, S.L. Miller of the College of Physicians and Surgeons, Columbia University, passed an electric spark through an atmosphere of gases such as the primitive atmosphere on earth would need to be composed of, if spontaneous generation could succeed. The apparent result was the formation of amino acids under these laboratory conditions.

Then, according to the theory of evolution, these tiny molecules, formed perhaps by an electric charge, found each other, clumped together, and interacted to form huge protein molecules large enough to bear life.

To the creationist this is interesting speculation, but is too far removed from probability to render it valuable. *If* by accident and chance the complex

Dr. Stanley Miller, whose experiment of passing a current through gases yielded several organic molecules. Formation of life, or imitation of life-like molecules in a test tube now seems a likely possibility. To get the molecules to reproduce is quite another matter.

proteins were formed is a big *if*.

But some evolutionists suggest that given enough time, *anything* can happen by chance. Dr. Harlow Shapley of Harvard University is quoted in *Science Newsletter* (July 3, 1965) as saying, "Life occurs automatically whenever the conditions are right. It will not only emerge but persist and evolve."

Yet men of equal ability doubt this. For instance, Malcom Dixon and E.C. Webb in their learned work on enzymes *(Enzymes,* second edition, New York: Academic Press, Inc. 1964, page 665) make this remark: "To say airily, as some do, that whenever conditions are suitable for life to exist, life will inevitably emerge, is to betray a complete ignorance of the problems involved."

Many of the scientists present at the symposium on the origin of life held at Moscow in August, 1957, felt that Oparin's idea that lifelike molecules could rise spontaneously was incredible. They could not and did not believe that large enough molecules of the right kinds of proteins could arise spontaneously to become the basis of organic life. Dr. Erwin Chartaff of Columbia University remarks, ''Our time is probably the first in which mythology has penetrated to the molecular level!'' (Erwin Chartaff, ''Nucleic Acids As Carriers of Biological Information,'' *The Origin of Life on the Earth,* pages 298-99)

The next step in the theory is that these large molecules somehow interacted to form ''simple'' cells. But we've already considered the complexity of the most simple! Millions of protein molecules would have had to ''spontaneously generate'' simultaneously not here and there around the world but in the same pond and at the same time. And then by strict accident and chance they would begin to ''count off'' and form themselves into various parts of the protozoa or other complex single-celled units.

Julian Huxley, born in 1887, is a British biologist and naturalist who has been the most ardent modern exponent of Darwinianism.

Does all this seem incredible? Yes, it does! But given enough eons of time, might not the impossible become possible? T.H. Huxley, a friend of Charles Darwin and grandfather of today's famous evolutionist Sir Julian Huxley, puts in stark relief the claim that *anything* can *eventually* happen by chance, a proposition the evolutionist can scarcely do without.

He declared that if a band of monkeys sat at typewriters and were given enough time—millions of years—it would be inevitable that at some time during this long period of time they would type the *Encyclopedia Brittanica* word for word and in exact order. Much of evolution requires this same degree of faith. Yet, until a

A major presupposition of evolution is that, given enough time, anything can happen by chance. The creationist doubts this, noting that chance can tear down as well as build up.

better theory is proposed, evolution
seems to many scientists to be too neat
a solution to many problems to be set
aside lightly. Especially since the
alternative—creation by God—is felt
somehow to be too "easy" a solution,
requiring too much faith.

That life generated spontaneously
is like saying that given enough time, a
house could build itself—that by chance
through millions of years, occasional
hurricanes would cause every board
and nail of a house to fit perfectly into
place, the plumbing to be installed,
the carpets to be carefully laid on the
floors. There are, however, two problems apart from the question of enough
time: First, will the lumber last, or
will it be ravaged by time, or knocked

106 apart again by one of the hurricanes? Secondly, will chance also produce the needed lumber mills, nail factories and carpet looms? Let us look for a possible method: a hunk of metal from a meteor lying in some field was picked up by a cyclone and worn flat to make a saw blade, then made jagged at the edges and transported to where other machinery, including electric motors and IBM computer equipment, was being assembled, all by chance and time.

This is an explanation of how a house might build itself. But it doesn't prove that it happened.

So also the explanation of molecular formation by Oparin and others does not prove that it happened that

way. The self-built house probably
involves fewer improbabilities than the
development of the first molecule, or
the first cell containing the fantastic,
incredible DNA apparatus and all the
other "plumbing and carpets"—enzymes,
mitotic apparatus, etc., etc.

No, a theory of how something
might have happened may have little
relationship to what *did* actually
happen. And as to the theory that
anything is possible by chance, given
enough time: zero possibility multi-
plied by infinity still yields zero.

The acceptance of this basic, car-
dinal presupposition of many evolutionists, that life generated spontaneously, must be questioned by thought-
ful high school students even though

it be presented to them by good
teachers and good textbooks. There
comes a time when a person must think
for himself.

   A further evidence of evolution,
to the evolutionist, are the so-called
vestigial organs. For instance the appendix, which, according to the evolutionist,
was used by man's ancestors who ate
much cellulose. Since cellulose would
require a much larger digestive organ
than modern man's stomach, the
appendix is supposed to have been a
large addition to the bowel system, now
no longer needed. But is this suggested
explanation a proof? No, merely a
theory that perhaps proves too much:
what shall we say of the vestigial breasts
and nipples possessed by every man
and other male mammal? Shall we
say that man once suckled his babies?

UNnatural Selection?

One of the beliefs of the evolutionist most difficult to accept is his assertion that tiny random mutations,
during millions of years, become complex new organs such as the eye and
ear. These incomplete changes would
often be detrimental to the animal
during millions of years until the final
organ was completed. Why then would
they be selected? Why would these
apparently pointless or unvaluable changes
be preserved by natural selection?

The evolutionist explains that each
tiny mutation which will produce prog-
ress is somehow beneficial, or at least
not lethal, and waits until another
mutation appears with which it can
cooperate. This is difficult to accept.

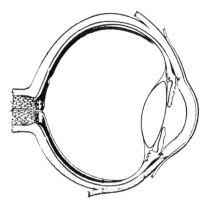

It would appear that here is a situation
where an explanation does not really
explain. It may be a necessity to help
the theory more than it is a fact.

As to the incredibly complex eye,
George L. Walls, in his book *The Verte-
brate Eye,* quotes Froriep as saying,
"It appears fully formed as though
sprung from the brow of Zeus." If so,
this is amazing—so amazing that the
creationist feels that his case is enor-
mously strengthened. For if the eye
didn't come by slow mutations, (and no
evolutionist is suggesting that it came
all at once as a major mutation), then
creation seems to be the logical
answer.

A perhaps even greater difficulty for the evolutionist are the male and female copulatory organs of mammals. For until these organs are complete, they are useless. And meanwhile, if the organs are not available for use, that kind of animal is doomed to quick extinction because it is without posterity.

The evolutionist may suggest possible transition organs during the eons until the present copulatory organs became functional, but once again there is no evidence that this was the case, or reason for incomplete copulatory organs to be selected by nature. Without doubt the evidence favors the creationist in this vast area of essential organs. Again let it be said that half-

formed organs would usually be in the
way rather than beneficial, and would
be passed over by natural selection,
rather than being chosen. And how
could the copulatory system of the
mammal be valuable until complete?
A half-formed male organ could not
produce and inject sperm into the female, nor could a half-formed egg
production system of the female be of
any value. Why then would it be
chosen? The better answer would seem
to be that these organs were not formed
by natural selection, but by creation
along with the entire species.

The question often rises among creationists as to the length of the "days" of creation mentioned in the Bible (Genesis 1). Were these 24-hour days? The answer lies in the fact that the Hebrew word translated "day" in the first chapter of Genesis can equally well be translated "period of time." How long is a "period of time"? The word is completely indefinite. It might refer to 24 hours and it might be millions of years. Apparently each day in Genesis was of millions of year's duration, if the normal system of dating fossils is accepted. In fact, snails appear in the fossil record 450,000,000 years ago! Man is perhaps up to a million years old! So we are talking about days 100,000,000 years long!

The "Family Tree of Man" according to the evolutionist. But to detail an idea is not to prove it. The factual evidence is largely missing.

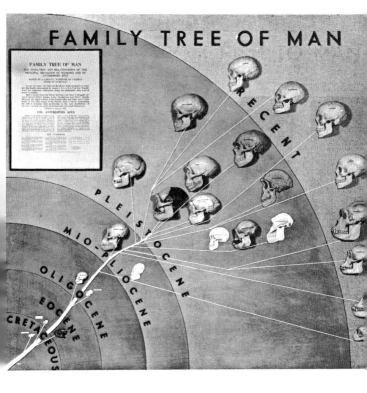

Another possibility urged by some creationists is that the days were 24 hours long, and all creation was finished in a 7-day week. To accept this idea requires believing that the fossil layers are meaningless; that they were planted by God at the time of creation rather than being laid down by geological action; or that the layers were all somehow formed in the year of Noah's flood.

To me it appears that God's special creative acts occurred many times during six long geological periods, capped by the creation of Adam and Eve perhaps a million or more years ago. This idea seems to do justice both to the Bible and to what geologists and anthropologists currently believe. If they change their dates up or down, it will make no difference to this belief, unless to move Adam's age forward or backwards.

"The Phylogeny of the Primates." Here again, the visual presentation helps us understand the theory, but the question is, did it happen that way? Other theories are equally credible and may be as easily assumed. Alert students who are not bound to biological traditions should be fully informed of alternative possibilities and be ready to examine them carefully.

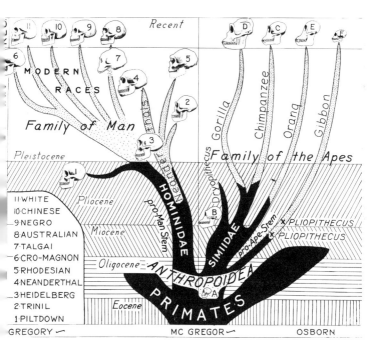

And now what about theistic evolution: the belief that God used evolutionary processes such as natural selection to bring a man's body into being, and then at some point God placed a spirit into that man-like body, so that it became a true man, Adam?

Well, if there is insufficient evidence for plain evolution, then there is also insufficient evidence for theistic evolution. But if you believe in evolution, can you still accept the Bible, or must you throw away your faith?

No, don't throw away your faith! The Bible account can be read to harmonize with current science, if you insist! To me this seems unwise, unnecessary, and wrong, but here is

the formula: Just assume that when I,
as a creationist, read the first chapter
of Genesis, I am reading into it something that isn't there. For I assume
that it says that the creation of man
was instantaneous. But when the Bible
says, "God created man in His own
image," perhaps this doesn't say or
mean "instantaneously," but means
instead that "God created man (by a
long process of evolution) in His own
image." You can argue that since the
Bible is not quite clear on this point,
it is not unBiblical to believe either
way you like. Among British evangeli-
cals this view is fairly common–that
God used evolution to form man. But
it is not my own view at all, as you will
know by now!

120 Remember, too, that the Bible's principal purpose is to reveal to all mankind the kindness and power of God to forgive our sins under the conditions the Bible sets forth. There is no controversy about this. Whatever you decide about evolution has no relationship to this far-reaching, utterly basic fact of sin and salvation.

The theory of evolution, although widely accepted, is based on many pre-suppositions. (One current, major text-book on evolution labels as assumptions twelve unprovable factors that are re-quired as articles of faith to make the theory work.) So the theory is very far from proved, but is a convenient explanation for those who find it diffi-cult to believe in instantaneous creation.

However, the theory of evolution is difficult to accept because it has so little evidence. The fossil record does not seem to correlate well with the theory because the fact is that each major kind of plant and animal appears suddenly without intermediate forms connecting it to earlier forms. To the creationist this means that God created these new forms from time to time without connection to previous forms. In each case God used the same cellular structure as part of His master plan, just as a builder uses lumber, windows and roofs for all sizes and styles of houses.

The theory of evolution also seems totally inadequate in solving the problem caused by another of its assumptions—that organs of the body evolved gradually through tiny mutations, selected by survival of the fittest. For these half-formed organs would often be detrimental, and would not be selected for survival. An eye, for instance, would hardly evolve slowly by chance mutations through millions of years. It requires a master architect. Male and female reproductive systems are another example.

Furthermore, the theory is entirely unreasonable in its assumption that such amazing units as cells, packed with chromosomes and other incredibly sophisticated machines like DNA, would develop by the chance organization of molecules, in turn formed by chance from inorganic material.

Conclusion

1. There is no hard evidence for evolution in the fossil records.

2. There is no sound theoretical basis for "upward" development, since the science of genetics as we know it does not permit variations apart from what is inherent in the original genes.

3. There is no imaginable way for chromosomes, genes, enzymes, DNA, etc., to have developed by chance and natural selection.

4. There is no evidence that "nature" has creative goals towards which it works for millions of years. This invests the blind forces of nature with foresight and personality. This concept of nature sounds like another name for God.

5. There is no way for complex organs to arise by minute and progressive mutations; natural selection would eliminate these useless pre-organs, instead of encouraging them.

6. There is no proof for the existence in nature of such a process as natural selection except for minor variations of size, color, facial feature, etc.